WILDLIFE
OF THE
CANADIAN ROCKIES

PHOTOGRAPHY and TEXT
by
WAYNE LYNCH

ASSISTED
by
AUBREY LANG

ALPINE BOOK PEDDLERS
CANMORE, ALBERTA, CANADA

Canadian Cataloguing in Publication Data
Lynch, Wayne
Wildlife of The Canadian Rockies
ISBN 0-9699368-0-X
1. Mammals—Rocky Mountains, Canadian (B.C. and Alta)—
Pictorial works. 2. Birds—Rocky Mountain, Canadian (B.C. and Alta.)—
Pictorial works. I. Lang. Aubrey. II. Title.
QI.221.R6L96 1995   599.09711   C95-910338-4

Design: Halle Flygare, Aubrey Lang, Wayne Lynch

Printed and bound in Canada by Friesens Books, Altona Manitoba.

Alpine Book Peddlers
Box 250, Canmore, Alberta Canada T0L 0M0

WILDLIFE of the CANADIAN ROCKIES is a specially commissioned portfolio of mammals and birds found in many of the national and provincial mountain parks of Alberta and British Columbia, including:

Jasper National Park

Banff National Park

Yoho National Park

Kootenay National Park

Waterton Lakes National Park

Bow Valley Provincial Park

Mt. Robson Provincial Park

Mt. Assiniboine Provincial Park

Peter Lougheed Provincial Park

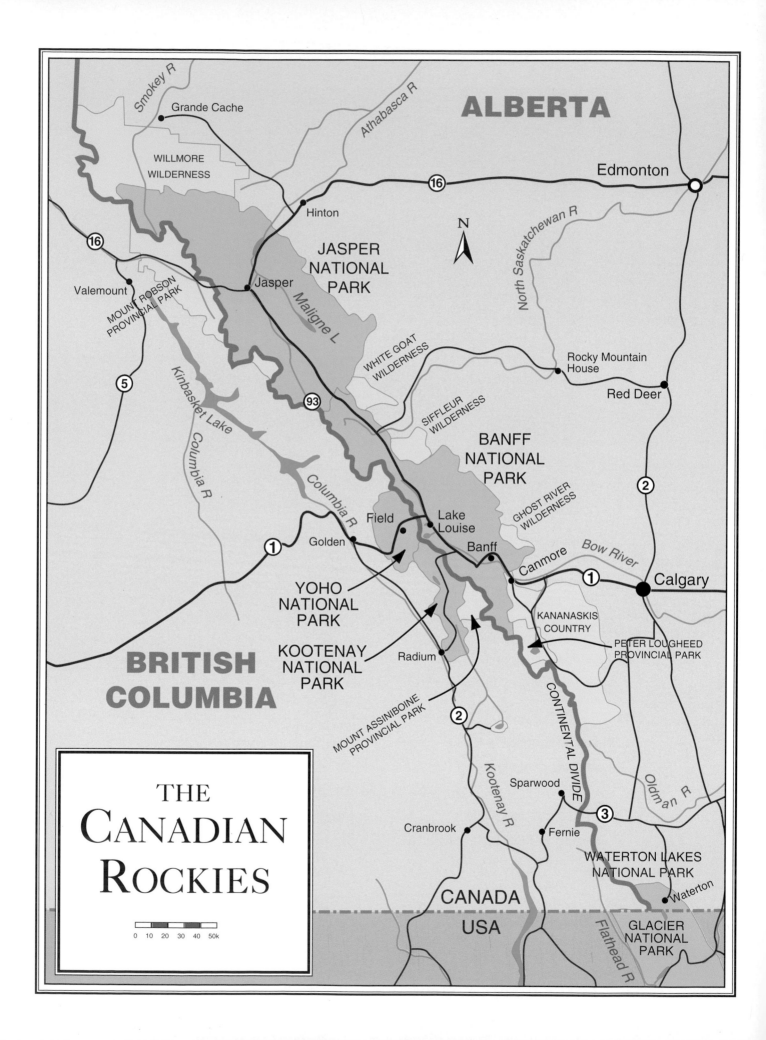

Smokey R
Grande Cache
WILLMORE WILDERNESS
Athabasca R
ALBERTA
16
Edmonton
Hinton
16
JASPER NATIONAL PARK
N
North Saskatchewan R
Valemount
MOUNT ROBSON PROVINCIAL PARK
Jasper
Maligne L
5
Kinbasket Lake
93
WHITE GOAT WILDERNESS
Rocky Mountain House
Red Deer
SIFFLEUR WILDERNESS
Columbia R
BANFF NATIONAL PARK
2
GHOST RIVER WILDERNESS
Field
Lake Louise
1
Golden
Banff
Bow River
Canmore
Calgary
1
YOHO NATIONAL PARK
KANANASKIS COUNTRY
PETER LOUGHEED PROVINCIAL PARK
KOOTENAY NATIONAL PARK
Radium
BRITISH COLUMBIA
CONTINENTAL DIVIDE
Oldman R
MOUNT ASSINIBOINE PROVINCIAL PARK
2
Kootenay R
Sparwood
3
THE CANADIAN ROCKIES
Cranbrook
Fernie
WATERTON LAKES NATIONAL PARK
0 10 20 30 40 50k
Waterton
CANADA
USA
Flathead R
GLACIER NATIONAL PARK

WILDLIFE of the CANADIAN ROCKIES

Dear Mountain Traveller:

From my home in Calgary, Alberta I can see the saw blade summits of the Rocky Mountains, only 100 kilometers (60 miles) away to the west. I see these mountains almost every day, and yet every time I catch sight of them they surprise me with their beauty, and I smile. At moments like these I often recall one of the special memories among the many I have of the mountains. I glimpse again the wet glistening body of an otter as it rolls in the water with a fish in its mouth. I remember a pack of four wolves as they lope across a forest clearing and melt into the twilight, or the rubbery lips of a black bear as it strips berries from a bush. My memories of the mountains always include wildlife. It is muscle and blood, feathers and fur that gives the mountains their life-force, and it is the wildlife which makes the mountains so compelling to visit and rewarding to explore.

The Rocky Mountains are the longest chain of mountains on earth, stretching 8500 kilometers (5100 miles) from the windswept tundra of northern Alaska to the desert plateaus of Mexico. The Rockies are their greatest, however, as they forge across the width of western Canada. Here, the mountains are a rugged spine of jagged peaks often cresting where eagles soar. With such length and height, the Canadian Rockies boast a varied range of habitats: icy summits, evergreen forests, flowered alpine meadows, aspen woodlands, rivers and lake shores, and grassy foothills.

The great diversity of mountain habitats sustains a rich array of wild creatures including nearly 280 species of bird, of which more than half are frequently seen. There are also 69 kinds of hairy mammals, 11 different croaking frogs and warty toads, and 3 kinds of slithering, silent snakes.

In a morning of hiking in the mountains, you can walk from Alberta to the Arctic. A climb of 300 meters (1000 feet) on the side of a mountain is roughly the same as a northward trek of 1600 kilometers (1000 miles).

The slopes of a mountain are layers of geography - aspen parklands stacked on grassy prairies, boreal forests on top of parklands, and alpine tundra on top of them all. Climbing from a valley bottom to above the treeline not only the vegetation changes but the climate does as well. Many animals in the mountains drift constantly between the different zones, some shift only as the seasons change, and some never stray for much of their lives. In the pages which follow, I have arranged the portfolio into three general habitats: foothills, wetlands, and high country. Naturally the divisions are somewhat artificial, and an animal placed in the wetland section, for example, may also be found in the high country, or the foothills, or in both. Generally, I tried to showcase the wildlife in the habitat, or habitats, most important to them.

Wildlife of the Canadian Rockies is my tribute to the magnificent and the mighty, the elegant and the elusive. In your travels through the mountains, I hope you will discover, as I have, the beauty and magic of its wild creatures.

Wayne Lynch

FOOTHILLS

Travelling west towards the mountains, the flat prairie landscape begins to gently roll and buck. Soon, the horizon disappears and the sky seems closer. Clumps of aspens cloak the hilltops and grassy meadows are mixed among the trees. This is foothill country, a 25 to 40 kilometer (15 to 24 mile) strip of terrain along the eastern flank of the Rocky Mountains in the "rain shadow" of the peaks. It's drier here than in the mountains, and the scantier rainfall encourages grasses to flourish. For wildlife, the foothills offer some of the luxuriant pastures of the prairies with the added protection offered by the shelter of the trees.

The foothills are a boundary zone between the horizontal prairie and the vertical mountains. As with any boundary area in Nature, the foothills attracts wildlife from the habitats on either side of it so there is a great diversity of mammals and birds.

Coyote

The coyote is one of the most widespread members of the wild dog family in North America. The animal's adaptability and flexible feeding habits enabled it to expand its range well beyond its original home in the central prairie grasslands of the continent where it lived in the early 1800s. Today, this cunning carnivore is found from Alaska to Panama.

Great Horned Owl

When young great horned owls leave the nest they are capable of short flights, but they depend upon their parents for food for most of the summer. This is a critical time for young owls – a time to hone their hunting skills and become independent.

Mountain Bluebird

The colour in a bird's feathers is produced either by pigments or by the optical effects of light on the microscopic structure of the feather. The intense blue of the mountain bluebird is an optical illusion caused by the structure of its feathers.

Western Meadowlark

The male western meadowlark has a repertoire of 6 to 9 different song types. Males with larger repertoires attract a mate sooner and produce a greater number of fledged offspring. It seems that repertoire size may be one way in which a female meadowlark can assess the quality of a prospective mate.

Nuttall's Cottontail

During the spring mating season, the docile-looking Nuttall's cottontail turns into a real fighter. The rabbits stand on their hind legs and cuff each other in the face and ears. Most fights were thought to involve males cuffing other males. It turns out that females do most of the clouting, and the rabbits they clout are courting males.

Cougar Kitten

In the Canadian Rocky Mountains, cougar kittens may be born in any month of the year, and indeed one-quarter of them are born during the cold winter months between October and March. The young cougar's spotted coat will gradually fade by the end of its first year.

Coyote Pups

Young coyotes are born in April or May, and there are usually 4 to 6 pups in a litter. Littermates begin to establish a pecking order very early in life and they may engage in serious fights with each other when they are only three weeks old. In this way, young coyotes quickly determine who is dominant, and who is not.

American Black Bear

In spring, although an adult black bear may sometimes stalk and kill newborn deer and elk, most of the time it behaves more like a cow than a carnivore and simply stuffs itself with green vegetation such as the leaves and flowers of dandelions.

Cinnamon Bear

The cinnamon bear is really just a brown-coloured American black bear. Less than 20 per cent of the black bears in the Rockies are varying shades of brown. These brown "black" bears are sometimes mistaken for grizzlies. The distinguishing characteristics of a grizzly are its prominent shoulder hump and much longer front claws.

Badger

The badger, a large member of the weasel family, has long claws to dig out tunnelling rodents such as ground squirrels and pocket gophers. The predatory badger also eats snakes, birds and insects. The badger is most active at night and spends the day curled inside its burrow.

Red Fox

A hunting red fox avoids competition with the other wild dogs of the Rocky Mountains, namely the wolf and the coyote, by specializing in mice and voles. The fox is a champion leaper. Its slender legs and lightweight frame allow it to leap up to 5 meters and pounce on small prey hidden in snow or grass.

Sharp-tailed Grouse
During spring courtship, the male sharp-tailed grouse inflates colourful air pouches in its neck and dances with other males on traditional stomping grounds. The dominant male grouse occupy the center of the dancing grounds and do most of the mating.

Richardson's Ground Squirrel
The Richardson's ground squirrel is well adapted to a burrowing life style. The animal's short front legs and long claws are efficient digging tools. Its small external ears keep dirt from getting trapped, and its highset eyes are well positioned to detect aerial predators.

Bobcat

As a bobcat travels throughout its home range, it frequently creates scent-marks using a combination of urine, feces and secretions from its anal glands. Small amounts of urine may be squirted on bushes, rocks, and snowbanks. Both urine and fecal scent-marks may also be used in association with scrapes made by the hind feet.

Deer Mouse

The deer mouse is a prolific breeder. A female has 2 to 4 litters a year, and each litter commonly contains 4 to 6 young. The young mice can start to breed when they are only 5 weeks old. It is no surprise then, that the deer mouse is so common and widespread.

Short-eared Owl

The short-eared owl hunts by gliding close to the ground, back and forth, across open grasslands and marshes. The large surface area of its long wings allows it to glide better than most owls.

Grizzly Bear

People often think that when a grizzly bear stands up, it is a sign of aggression and the bear is about to charge. Actually, when a grizzly stands on its hind legs it is only trying to get a better view of the surroundings to decide what it should do next.

Bison

During the last, century when millions of bison roamed the central prairies of the continent, the animals often migrated to the shelter of the foothills for the winter. Today, all of the bison found in the Rocky Mountains are in captive herds.

Wolf Pup

Pelt colour in the wolf varies considerably, from white to gray, and black. Other colour variations include brown, tan, and buff. Certain colours tend to predominate in a given geographic location. Black and gray-coloured wolves are common in the Rocky Mountains.

Cougar

Cougars living in mountainous regions, where there is heavy snowfall in winter, shift from higher elevations in summer to low elevations in winter. The cougars simply follow the seasonal movements of the elk, deer, and moose on which they prey. There is a healthy population of cougars living in the foothills of the Canadian Rockies.

Bobcat

The bobcat is the smallest of the three wild cats found in the Canadian Rocky Mountains. It is about twice the size of a domestic cat but is more muscular and weighs up to 12 kilograms (26 pounds). Male bobcats are approximately a third larger than females.

Porcupine

The porcupine is primarily a vegetarian and in winter feeds almost exclusively on the bark of trees and woody shrubs. In summer, it shifts from the treetops to the ground where it nibbles and chews on green leafy vegetation. The porcupine's armour of 30,000 barbed quills protects it from predators while it shuffles about on the ground.

Mule Deer

A mule deer buck is in its prime at the beginning of the annual rut in late October and early November. At this time, its antlers are polished clean, its neck is swollen with muscle, and its fat reserves are at their peak.

Red Fox Pup

In the northern Rockies, a litter of red fox pups may contain some foxes with the usual red fur, some with black fur which are called "silver" foxes, and some with the coloration pictured which is called a "cross" fox. The cross fox gets its name from the distinct cross of dark fur across the animal's shoulders.

Mule Deer

The mule deer is found in badlands, deserts, grasslands and mountains throughout the western half of the continent from the Yukon to northern Mexico. The largest individuals, however, occur in the Rocky Mountains. Females, such as the one pictured, average 59 kilograms (130 pounds) and males weigh an average 74 kilograms (163 pounds).

Cougar

Popular belief has it that a hunting cougar hides beside a game trail and then pounces on any unwary prey which happens by. Actually, a cougar stealthily stalks to within 15 meters (49 feet) or less and then leaps on to its victim after a few bounding strides. If the cat can stalk to within striking distance it will be successful in over 80 per cent of attempts.

WETLANDS

A single drop of meltwater drips from the lip of a glacier high among the peaks of the Rockies. Soon the droplet becomes a trickle, then a stream, and finally a tumbling river empties into a quiet lake, tinted like polished jade. At the far end of the lake the droplets are carried into other rivers, and from there into marshes, lakes, and still larger rivers until finally the mountains are left behind.

In the journey from mountain peaks to valley lowlands the waterways and wetlands of the Rocky Mountains have many different faces. There are silted rivers where eagles, fish and ducks compete, and vast emerald lakes haunted by loons. There are marshlands rimmed with rushes, and beaver ponds where mink and moose abound. Look for water and you will find life.

Beaver

The beaver is the largest rodent in North America, and an adult beaver may weigh as much as 31 kilograms (68 pounds). The bark-eating beaver stakes its whole existence on two pairs of prominent incisor teeth. As if to highlight the importance of these teeth, they are bright orange in colour. A beaver's incisors grow continuously throughout the animal's life, and are self sharpening as the upper pair grinds against the lower one.

Muskrats

The muskrat is very flexible in where it lives. In marshlands the animal frequently builds a lodge consisting of a piled-up mound of cattails, bulrushes and other aquatic vegetation. Along rivers and streams in the mountains it may simply dig a burrow into the banks of the waterway. This pair of muskrats were grooming and snoozing in the sunshine on the edge of their lodge in a marsh.

Yellow-headed Blackbird

The yellow-headed blackbird nests in loose colonies in marshlands, especially where there are extensive areas of dense bulrushes. Normally the yellowhead is dominant over the more common red-winged blackbird. If preferred habitat is at stake, the yellowhead will oust the redwing.

Red-winged Blackbird

The scarlet epaulets of the male red-winged blackbird are a badge of ownership and are displayed most conspicuously whenever the bird sings its familiar territorial song. When scientists blackened the red shoulder patches on a group of males, more than half of them immediately lost their territories to intruders.

Canada Goose

No bird of the Rocky Mountains is more associated with the passage of the seasons than is the familiar Canada goose. In spring, the honking of geese overhead forecasts the end of winter, and great flocks of Canadas wedge their way south again as the last golden leaves of autumn flutter to the ground.

Bull Elk

The elk pictured is a "six-point" bull, so named because of the number of tines it has on each of its antlers. A pair of such antlers might weigh 13 kilograms (29 pounds). The bull will shed its antlers in late March, almost five months after the autumn rut is over. A new set of antlers begins to grow again by the end of May.

Canada Geese

When families of Canada geese are raised near one another, the goslings may combine into groups, called creches. As a result, several dozen goslings of assorted ages may be under the care of a single adult pair. The record is 68 goslings with one pair of adult Canadas.

Common Goldeneye

Like many diving ducks, the common goldeneye can dive to depths of 7 meters (22 feet) in search of food. This is shallow in comparison to some of the other diving birds such as the common loon which has been recorded at depths of over 66 meters (200 feet).

Hooded Merganser

Not surprisingly, courtship behaviour in the male hooded merganser is marked by crest erection, head-shaking, and head-throw displays, all of which serve to emphasize the duck's large crest.

Bald Eagle

The bald eagle has the largest wingspan of any bird of prey found in the Rocky Mountains. Adult female eagles are as much as 20 per cent larger than adult males, and the wings of a large female may span well over 2 meters (6½ feet).

Bald Eagle

The bald eagle has a long life span, as much as 50 years in captivity and sometimes over 30 years in the wild. As a result, an eagle will not acquire the full characteristic plumage of an adult until it is 4 or 5 years old. The eagle pictured still has a few dark juvenile feathers on its head.

Horned Grebe

Although this bird looks like a duck, it belongs to a completely separate group of diving birds called grebes. The bird is a horned grebe and it is incubating a clutch of 5 eggs on a floating nest of aquatic vegetation which it pulled up from the bottom of the lake.

Yellow Warbler

The smaller the bird, the more we marvel at its feats of endurance. The yellow warbler, weighing only a few grams (less than an ounce), may fly more than 2700 kilometers (1600 miles) in its autumn migration to Mexico. For a human to achieve an equivalent feat would require travelling the distance from the earth to the moon six times.

White-Crowned Sparrow

The white-crowned sparrow will eat the catkins of willows, as well as mosquitoes, flies, caterpillars, spiders, and beetles. Its main food, however, is the seeds of weeds and grasses as you might have predicted from its strong, cone-shaped bill.

Moose

Mammals need salt throughout the year, but the period of greatest salt hunger seems to occur between May and August. Moose satisfy their salt cravings at this time of the year by eating aquatic plants which contain 300 times more salt than the dry twigs which the animals eat in winter.

Red-Necked Grebe

In common with all grebes, the red-necked grebe has a habit of eating its own feathers. Even young grebes, a few days old, are fed feathers by their parents. It is speculated that the feathers enmesh swallowed fish bones and prevent injury to the bird.

Osprey

The osprey is unique among birds of prey in having a reversible outer toe on each of its feet. Sharp spicules also cover the bottom of the bird's feet. Both features are adaptations which facilitate the capture of agile, slippery fish, the osprey's principal prey.

Common Merganser

The common merganser derives its unusual name from two Latin words: *mergere* meaning to plunge or dive, and *anser* meaning goose. The merganser is well named as it is an expert diver which swims swiftly underwater in pursuit of small fish, its main food.

Muskrats

The muskrat's name refers to its musky odour during the breeding season. Both sexes have scent glands which enlarge during the mating period, but the glands are most active in males.

Harlequin Duck

The harlequin is the characteristic duck of fast-flowing mountain rivers and streams. The colorful birds are completely at home in white water where they search for aquatic insects.

Red-necked Grebe
The legs on a red-necked grebe are positioned farther rearward than in any duck. The grebe's feathers also trap less air, its bones are more solid, and its air sacs are smaller - features which make the grebe less buoyant than ducks, and therefore a better diver.

Cinnamon Teal
The handsome male cinnamon teal is the only all-reddish duck in the Canadian Rockies. It is a small duck which dabbles along the edge of lakes and ponds searching for seeds, insects and snails.

Red-tailed Hawk

The red-tailed hawk forms a lifelong pair-bond with its mate. The monogamous bond is maintained until the death of a partner. At all times of the year, redtails are seldom out of each other's sight.

Great Blue Heron
The long, specialized feathers on the breast of the great blue heron are called filoplumes, and are part of the bird's breeding plumage.

Bull Moose

The moose is the largest member of the deer family. A large bull may weigh more than 450 kilograms (1000 pounds) and have a shoulder height of over 2 meters (6½ feet). As in all deer in North America, male moose are substantially larger than females.

Bald Eagle

A bald eagle will feed on carrion whenever the opportunity arises. This bird has found a dead fish. When feeding, an eagle tears off large chunks of flesh and swallows them whole. The eagle eats quickly since there is always the risk that another eagle may fly by and pirate the meal.

Common Loon

The common loon is the quintessence of wilderness. The Cree believed that the loon's plaintive yodel was the cry of slain warriors calling back to the land of the living.

American Black Bear

The black bear is a strong swimmer and will readily swim to islands in search of food. The bear also travels to water in hot weather to wallow and cool off.

Tree Swallows

The female tree swallow, pictured on the right, is the nest builder. She chooses a natural cavity or an abandoned woodpecker hole, oftentimes near water. If nest sites are scarce, the tree swallow will readily accept a bird box.

Mallard

Minimizing the body heat lost through its legs and feet is very important to the mallard's survival, as it is to all birds that stand on ice or swim in cold water. The blood vessels in a bird's legs are arranged so that the artery is completely surrounded by multibranched veins. This arrangement acts as a heat exchanger and reduces heat loss.

Bufflehead

The name bufflehead is a contraction of buffalo-head. Early travellers thought the duck's head was too big for its body and was shaped like the head of a buffalo. This handsome little duck is a delight to watch anytime, but especially during the breeding season when males court females in wild aerial chases.

HIGH COUNTRY

Along the length of the Canadian Rockies a great evergreen forest of spruce and fir cloaks the slopes of the mountains from valley bottoms to timberline. This is possibly the largest habitat in the Rockies, yet its use by wildlife is less than you would expect. The shadowed forest of conifers offers shelter and refuge, but there is relatively little for wildlife to eat. The exceptions to this are avalanche slopes where tons of snow have bulldozed down a mountain side and mowed a path through the trees creating an oasis of open space. In the aftermath of the slide, edible shrubbery and berry bushes grow in profusion and animals and birds are lured to the area.

The upper boundary of the forest is the timberline which hovers around 2000 meters (6500 feet). At these altitudes, the last of the trees disappears. Here at the top, the winters are long and cold, the winds are high, and it may snow on any day of the year, making it a challenge for the wildlife as well as the plants.

Above the treeline the alpine tundra is barely distinguishable from the Arctic tundra. Drawn by the uncrowded conditions, the birds and mammals which inhabit these treeless expanses do so mostly in summer, but most will flee to lower altitudes as soon as winter returns.

Lynx

The lynx is a snowshoe hare specialist. The cat's large feet enable it to run more easily across the snow when chasing its fleet-footed prey. The survival of the lynx is so dependent upon hares that lynx numbers mirror exactly the hare's ten-year cycle of boom and bust.

Woodland Caribou

The woodland caribou is the only member of the deer family in which the females (pictured), as well as the males, sport a rack of antlers. It is a mystery why this should be.

Steller's Jay

In the western third of North America, Steller's jay replaces the more familiar blue jay. The white eyebrow line distinguishes the mountain subspecies from the coastal subspecies in which the eyebrow line is absent.

Pine Grosbeak

Some birds, such as the beautiful pine grosbeak, migrate vertically instead of horizontally and still reap the benefits of long-distance migration without having to log the miles. In the Rocky Mountains, the pine grosbeak nests at high altitudes in summer and then descends to lower levels in winter.

Gray Jay

Every spring, the gray jay is one of the earliest birds to nest in the Canadian Rockies. The jay builds a bulky nest of twigs, grass and bark strips in March, at a time when temperatures still plummet well below freezing, and snow still covers the ground.

Grizzly Bear
Grizzlies move into the high country in October to dig a den for the winter. The bears in the Canadian Rockies may hibernate for more than six months out of the year, emerging from their dens to face the spring sunshine in April and early May.

Wolf
Predators have a new image these days, and in many areas their numbers are being allowed to increase. A good example of this is the wolf in the national parks of the Canadian Rockies. The last wolf "control" program was carried out in the 1950s. Today, wolf packs once again roam the valleys of the mountain parks and wolf sightings are relatively frequent.

Great Gray Owl

Bird-watching is one of the fastest growing hobbies in North America with bird field guides currently outselling bibles. Among bird-watchers, the secretive great gray owl is one of the ten most sought after birds in North America.

Blue Grouse

The fleshy, yellow comb above the eye of the male blue grouse, as well as the bird's crimson throat patches, are most visible during the spring breeding season. A displaying grouse distends his throat with air causing the feathers to separate and expose the colourful underlying bare skin like a flower blooming.

Northern Saw-whet Owl

The northern saw-whet is a small owl only 19 centimeters (8 inches) tall. During the day, the saw-whet hides in the thick branches of a tree and is quite tame if discovered in its hiding spot. The owls are so unwary that researchers can sometimes catch the birds with their hands.

Grizzly Bear

The front claws on a grizzly bear are one of the animal's most impressive features. The claws vary in colour from black to ivory, with dark brown being the most common colour. Typically, the front claws are almost twice as long as the rear claws and may be 9 centimeters (3 inches) long.

Bull Elk

Bull elk bugle during the autumn rutting season to advertise their vigour and their whereabouts. A high ranking bull may bugle as often as 48 times in 30 minutes, and the more vigourous a bull bugles the more females he can attract to his harem with which to breed.

Bohemian Waxwing

In winter, the bohemian waxwing frequently travels in large flocks which drift around like bands of nomads. The name waxwing is derived from the bright red, waxlike material that forms on the tips of the wing feathers. The function of these droplets is unknown.

Ruffed Grouse

The thumping noise of a drumming ruffed grouse is audible to the human ear up to a half kilometer (¼ mile) or more. It would seem easy then for one of the bird's major enemies, the great horned owl, to locate and capture the drumming grouse. It turns out, however, that the frequency of the grouse's drumming sounds, those easily heard by us, are inaudible to the owl.

Spruce Grouse

The first bird-watchers in North America were the Indians, and the names they gave to birds showed that they were keen observers of bird behaviour. The spruce grouse was called the "bird that picks at the buds of evergreens and weeps," a reference to its eating habits and the crimson combs above its eyes.

Golden Eagle

In 1992 a previously unknown golden eagle migration route was discovered in the Front Range of the Canadian Rocky Mountains. Each year, over 4000 golden eagles, as well as hundreds of bald eagles, hawks, and falcons use the route in spring and then again in autumn. In 1994, over 800 golden eagles migrated through in a single day!

Golden-mantled Ground Squirrel

The golden-mantled ground squirrel lives at the highest altitude of the three ground squirrel species which live in the Rockies. The golden-mantle often lives at the very edge of the treeline and sometimes in the alpine tundra.

Peregrine Falcon
The speed of the peregrine falcon's power dive, or "stoop" as it is called by falconers, has been estimated at over 300 kilometers (180 miles) per hour.

Bighorn Sheep

Bighorn sheep spend much of their time feeding in open meadows and retreat to rocky cliffs only if danger threatens. As further protection against predators, mountain sheep travel in bands, composed of either adult ewes with juveniles and lambs, or all male bachelor herds.

Mountain Goat

The mountain goat, unlike the bighorn sheep, is a cliff specialist and the goat spends much of its time clinging to narrow ledges, leaping across deep chasms, and executing heart-stopping climbing manoeuvers. The prominent front shoulders of the mountain goat are heavily muscled to enable the goat to climb steep cliff walls with apparent ease.

Bighorn Sheep

By the time the horns on a bighorn reach this size the ram is at least 8 years old. The tips of the horns have become roughened and worn, called "brooming". In most cases, brooming is the result of fights during the breeding season when rival rams batter each other in round after round of head-bashing.

Clark's Nutcracker

Clark's nutcracker lives in the subalpine zone of the Rockies. In times of food surplus the nutcracker buries seeds, acorns, and juniper berries under the litter of the forest floor. Then, when food is scarce, the bird relocates its caches and raids them. The nutcracker has the remarkable ability to remember where it buried food, even when the location is covered under 15 centimeters (6 inches) of snow.

Hoary Marmots

The hoary marmot is the largest squirrel in North America, weighing as much as 9 kilograms (20 pounds) when it is padded with fat at the beginning of hibernation. The marmot's large size is an adaptation to the cold temperatures and protracted winters the animal must endure in its life at high altitudes. As a result of the severe climate, a hoary marmot may spend two-thirds of its life in hibernation.

Ravens

Ravens are playful birds. Researchers have even watched ravens play with grizzlies bears. Most often, however, ravens play with others of their own kind. The ravens pictured were gliding back and forth in the updrafts across a cliff face. During one of the passes, one of the birds rolled on its back and tried to playfully grab its companion with its feet.

Bighorn Sheep

Horns are different than antlers. Horns, such as those on this bighorn ram, are composed of a bony core covered by a keratin sheath. Keratin is the same material found in fingernails. Antlers, in contrast, are composed of solid bone. Whereas horns are retained throughout the life of the animal, antlers are shed and replaced each year.

Northern Hawk Owl

The top of a broken snag is the most common place for a northern hawk owl to nest. The three young owlets in this nest were being brooded by their mother against the cold winds of spring.

Black Bear Cub

Bears are the only mammals known to give birth while they hibernate, and thus bear cubs are born in the middle of winter. A newborn bear cub is no larger than a chipmunk but by the time the cubs leave the den in spring they weigh 3 to 4 kilograms (6 to 9 kilograms).

Bull Elk

The elk is probably the most common mammal seen by visitors. The name is confusing for Europeans since in Europe they call the moose an elk, and they call the elk a red deer. The naming game gets even more complicated when you consider that the Shawnee native name for the elk is wapiti, meaning "white rump", and this name has also crept into popular usage.

Pika

The tiny pika lives among the cracks and crevasses of rocky slopes. Although the pika looks mouse-like, it is not a rodent at all, but in fact is related to rabbits and hares. This explains why the pika's other common name is "rock rabbit".

Boreal Owl

The boreal owl, like all owls, has a number of adaptations which make the bird a better hunter: front facing eyes which improves depth perception, serrated wing feathers which reduces air turbulence and silences flight, and strongly taloned feet.

Lynx

When a lynx is at rest, its whiskers are relaxed and project to the sides. When it is capturing prey, however, the whiskers are extended forward like a net in front of the mouth so that the lynx can judge exactly where the prey is for an accurate killing bite.

FURTHER READING

- *Bears: Monarchs of the Northern Wilderness,* Wayne Lynch, Greystone Books, 1993

- *A Beast the Color of Winter,* Douglas Chadwick, Sierra Club Books, 1983

- *Mountain Sheep: A Study in Behavior and Evolution,* Valerius Geist, University of Chicago Press, 1971

- *Hoofed Mammals of Alberta,* J. Brad Stelfox, Lone Pine 1993

- *Cougar: The American Lion,* Kevin Hansen, Northword, 1992

- *Wolf Story, From Varmint to Favourite,* Dick Dekker, BST Publications, 1994

- *Birder's Handbook: A Field Guide to the Natural History of North American Birds,* Paul Erlich, David Dobkin & Darrly Wheye, Simin & Schuster, 1988.

- *Handbook of the Canadian Rockies,* Ben Gadd, Corax Press, 1986